FLORA OF TROPICAL EAST AFRICA

JUNCAGINACEAE

D. M. Napper

Annual or perennial herbs. Rhizomes short or long, or base bulbous; stolons rarely present. Leaves basal, linear, flat or semiterete. Inflorescence a spike or raceme. Flowers regular and bisexual, sometimes imperfect or unisexual. Perianth of 6 similar segments in 2 whorls, deciduous, herbaceous; tepals free or connate below. Stamens 6, rarely 4, in 2 series opposite the tepals; filaments usually short; anthers 2-thecous, dehiscing longitudinally, extrorse. Carpels superior, free or axially connate, 6, or 3 + 3 undeveloped, unilocular; ovules 1 or 2, collateral, basal, erect; stigmas sessile or subsessile, plumose or papillose. Mericarps 3–6, indehiscent, free or axially connate.

A small but widespread family of swamp and marsh plants with three genera, of which only *Triglochin* L. is known to occur in Africa.

TRIGLOCHIN

L., Sp. Pl.: 338 (1753) & Gen. Pl., ed. 5: 157 (1754)

Small herbs with naked scapose stems. Leaves graminaceous, flat or semiterete, with a sheathing base slit anteriorly, often ligulate at junction of sheath and blade. Inflorescence a single compact raceme or spike elongating in fruit. Flowers regular, bisexual; partially aborted flowers not uncommon in most species. Stamens 6, the inner whorl sometimes reduced. Carpels 6, alternate ones sterile; ovules solitary; stigmas 3, plumose. Mericarps 3, becoming free at maturity or axially connate, 1-seeded.

A small genus of about 15 inconspicuous marsh or swamp herbs, most of which occur in the temperate regions of the northern and southern hemispheres, with a few recorded in tropical and subtropical regions.

T. milnei *Horn af Rantzien* in Svensk. Bot. Tidskr. 55: 85 (1961). Type: Zambia, Mwinilunga District, *Milne-Redhead* 3012 (K, holo.!, S, iso.)

Slender bulbiferous perennial herb 15–45 cm. high. Bulbs solitary or clustered, surrounded by dense layers of fibrous leaf-base remnants. Leaves linear, 3–25 cm. long, sheaths 4–6 mm. wide, the blades narrower, glabrous. Flowers numerous, shortly pedicellate, 2·5–5 mm. long, green, the perianth and stigmas tinged with purple; pedicel 1–5 mm. long, elongating to 3–10 mm. in fruit. Outer tepals ovate, 2·5–4 mm. long, broadly acute, entire; inner tepals shorter and more obtuse, sometimes with the upper margin irregularly toothed. Outer anthers 1·5–3 mm. long, the inner ones scarcely more than half this size. Ovary 2·5 mm. long, with 3 free stigmas. Fruit narrowly lanceolate to elliptic, 8·5–14 mm. long, up to 3·75 mm. broad, of 3 fertile mericarps alternating with 3 much reduced; fertile mericarps lanceolate, narrowing above into a recurved tip, separating from the central axis at maturity. Seeds semi-elliptic, compressed, 4–6·5 mm. long, ± 1 mm. wide. Fig. 1.

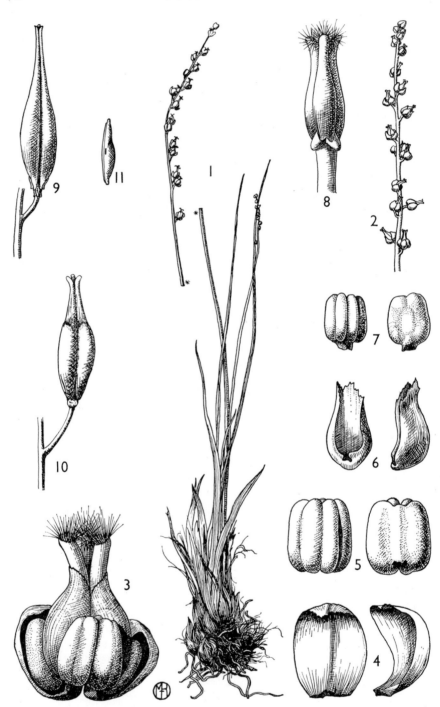

FIG. 1. *TRIGLOCHIN MILNEI*—**1**, plant, × ⅔; **2**, raceme, × 1; **3**, flower, × 10; **4**, outer tepal, × 10; **5**, anther of outer stamen, × 10; **6**, inner tepal, × 10; **7**, anther of inner stamen, × 10; **8**, gynoecium, × 10; **9, 10**, fruits, showing variation in shape, × 3; **11**, seed, × 3. 1–8, from *Milne-Redhead* 3012; 9, 11, from *Milne-Redhead* 3693; 10, from *Bullock* 2364.

TANGANYIKA.   Ufipa District: Sumbawanga, 30 Jan. 1950, *Bullock* 2364!; Iringa
   District: 6·5 km. N. of Iringa, 5 Feb. 1962, *Polhill & Paulo* 1362!; Songea District:
   Nonganonga stream 12 km. E. of Songea, 28 Dec. 1955, *Milne-Redhead & Taylor* 7934!
DISTR.   T4, 7, 8; Zambia, Rhodesia and Angola to South Africa (Natal)*
HAB.   Marshes and seasonally flooded grasslands; 1000–1900 m.

SYN.   [*T. bulbosa* sensu C.B. Cl. in F.T.A. 8: 215 (1902), as "bulbosum", *non* L.]

NOTE.   The gatherings from **T4** and **T7** have shorter, broader, almost elliptic fruits
   (fig. 1/10), compared with the narrowly lanceolate Zambian gatherings (fig. 1/9),
   containing similar but slightly broader seeds.   That the only gatherings from this part
   of the Flora area should both represent the same extreme form of a variable population
   is unlikely, but until further material becomes available it is not possible to propose
   with confidence that these gatherings represent a separate subspecies.

* The sole South African record given by Horn af Rantzien (*loc. cit.*) is doubtful and
may well be *T. bulbosa* L. The circumscription of *T. bulbosa* in F.S.A. 1: 93 (1966), which
includes *T. milnei*, is less satisfactory.

INDEX TO JUNCAGINACEAE